Software Reviews & Audits

A How To Guide for Project Staff

David Tuffley

To my beloved Nation of Four
Concordia Domi – Foris Pax

*Reviews and audits are an indispensable part
of any software project, though often omitted.*

Acknowledgements

I am indebted to the Institute of Electrical and Electronics Engineers on whose work
I base this book, specifically IEEE Std 1028.

I also acknowledge the *Turrbal* and *Jagera* indigenous peoples, on whose ancestral
land I write this book.

Contents

Contents

Contents

A. Introduction

This standard encompasses the range of review and audit activities undertaken during a project.

Reviews and audits comprise the following five modules.

- **Management review** - the formal evaluation of a project level plan or project status relative to that plan by a designated review team.

- **Technical review** - the evaluation of specified software modules and documents aimed at ensuring that the software modules and documents comply with the applicable standards while conforming to the specifications. The technical review may provide recommendations after the examination of alternatives.

- **Walkthrough** - an evaluation process that can result in recommendations for improvement or identification of alternatives to the current software modules or documents being developed.

- **Audit** - provide objective evidence of compliance of products and processes with standards, guidelines, specifications and procedures. Includes audits of the quality management system.

- **Inspection** - rigorous formal evaluations designed to detect and identify defects in the reviewed material. Normally conducted after the event and initiated by persons outside of the project team.

Reviews and audits support the overall quality objectives of QMS. These objectives include the need for evaluation, verification, and validation and compliance confirmation.

A.1. Scope

The scope of this standard applies to all project documents, software conduct specific reviews or audits - that need is defined by the project and quality plans. Also applies to the audit of the quality management system.

A.2. Objectives

The objective of this standard is to provide definitions and uniform requirements to allow project staff to perform the necessary reviews and audits of products and processes.

A.3. References & sources

[1] IEEE Std 1028-1988 - *Standard for Software Reviews and Audits*,IEEE Software Engineering Standards Collection, Spring 1991

A.4. Definitions & acronyms

Audit - an independent evaluation of software products or processes to ascertain compliance to standards, guidelines, specifications, and procedures based on objective criteria that specify:

(1) The form or content of the products to be produced
(2) The process by which the products shall be produced
(3) How compliance to standards or guidelines shall be measured. (AS4009)

Auditor - the person, with the required training, who carries out the quality audit.

Auditee - the person representing the project, process or company being audited.

Inspection - a static analysis technique that relies on visual examination of development products to detect errors, violations of development standards, and other problems. e.g. code inspection, design inspection. (IEEE 610.12-1990)

Issue/Defect - a non-conformance to a standard, a specification or a recognised method. An issue is a non-conformance that is found through verification while the deliverable is still under development. A Defect is a non-conformance that is detected during validation, that is, after the deliverable has been developed.

Major Defect - defects that would result in failure of the software module(s) or an observable departure from specification.

Management Review - a formal evaluation of a project level plan or project status relative to that plan, by a designated review team. (AS 4009-1992)

Minor Defect - defects that would affect only the non-functional aspects of the software module(s).

Review - an evaluation of software module(s) or project status to ascertain discrepancies from planned results and to recommend improvement. This evaluation follows a formal process (for example, management review process, technical review process, software inspection process, or walkthrough process). (AS 4009-1992).

Software Module - a deliverable or in-process document produced or acquired during software development or maintenance. (AS 4009-1992) This covers planning documents, specifications, reports, other documentation and source code.

Technical Review - a formal team evaluation of a software module(s). It identifies any discrepancies from specifications and standards or provides recommendations after the examination of alternatives, or both. (AS 4009-1992).

Verification - confirmation that the prescribed work has been done.

Validation - confirmation that the prescribed work has been done correctly.

Walkthrough - a static analysis technique in which a designer or programmer leads members of the development team and other interested parties through a segment of documentation or code, and the participants ask questions and make comments about possible errors, violation of development standards and other problems. (IEEE 610.12-1990)

A.5. Responsibilities

In addition to the following general responsibilities, certain review/audit-specific responsibilities also apply. Details of these are found in the staffing section of the relevant review or audit.

The Project Manager is responsible for the following:

- Co-ordinating reviews and the anticipated rework.

- Conducting the review and reporting the results against the project plan.

- Providing the necessary resources of time, personnel, budget and facilities needed to plan, define, execute and manage the reviews.

- Track status of reviews.

- Ensuring that the deficiencies identified in the review and audit reports are rectified.

The Quality Manager is responsible for the following:

- Ensuring that quality audits are scheduled and carried out.

- Ensuring the required level of reporting and follow-up action is carried out.

- Ensuring that quality records are maintained, covering:
  - The audit schedule
  - The audit register
  - Audit documentation

- Ensuring that staff nominated as auditors are independent of the area being audited.

- Ensuring that staff nominated as auditors are suitably qualified and briefed for the audit.

- Arranging suitable training for staff nominated as auditors.

Development staff are responsible for the following.

- Having a level of development expertise and product knowledge sufficient to understand the software being reviewed.

- Production of documents and software items to comply with the relevant standards and specifications.

- Participation in the review process.

B. Software Reviews & Audits

The sections which follow give a detailed outline of the procedures for each review/audit type (i.e. management and technical reviews, inspections, walkthroughs and audits).

B.1. Overview

B.1.1. When to do what

Figure 1 below illustrates the way in which the various reviews and audits are used and how they differ from each other. The processes have two main distinctions, as follows.

- **Subject matter** - management reviews and audits evaluate the project as a whole, whereas walkthroughs, technical reviews and inspections are applied to individual documents or software modules.

- **Means of initiation** - Audits and inspections are normally initiated by parties external to the project team, such as the quality manager or the steering committee. They are done by people selected by the initiating party who are independent of the project team, though the project team would still assist with information. Walkthroughs and technical reviews, and to a lesser extent management reviews are common project activities initiated by the project team during

10

the normal course of a project. Walkthroughs are an optional review which are performed prior to completion of the document or software module.

B.1.2. Review status register

A review status register shall be created to record the progress of reviews. The register shall record at least the following details for each review performed.

- Review report document number.

- Main reviewed item document number.

- Review start date.

- Review end date.

- Number of decisions.

- Number of action points.

- Number of major issues/defects.

- Review effort.

- Size of reviewed material.

- Exit decision (pass, provisional pass or fail).

- Date action points were completed.

- Date all issues/defects were corrected.

- Review outcomes finalised date.

- Status (planned, active or complete).

B.1.3. Mandatory and optional practice

Mandatory - in the conduct of Quality Management System (QMS) projects the following activities are mandatory.

- Technical review of all documents prior to approval.

- Management reviews of projects at the completion of each project phase.

Optional (recommended) -in the conduct of QMS projects the following activities are optional but recommended.

- Walkthroughs of documents during their development as a means of improving document quality.

- Technical reviews of selected software modules to ensure standard and specification compliance.

Optional - in the conduct of QMS projects the following activities are optional only.

- Walkthroughs of software modules to identify areas for improvement.

- Audit(s) of the project to establish compliance with QMS requirements.

- Inspections of documents and software modules to verify compliance with QMS standards.

- Technical reviews of all software modules.

B.1.4. Partial or complete reviews

Reviews are designated in the objectives as being either partial or complete reviews. Partial reviews examine only those elements which have changed. Partial reviews can only be done after a complete review.

When a document, software item or other project item is developed, it undergoes a review (technical review) to validate its content, and compliance with the applicable standards. (For documents, approval can only occur after the review and any follow-up actions have been resolved.)

This review must be a full review of the item. That is it must review the entire content of the item. If the item is later modified in some way another review must be performed. This review may elect to cover only the changed portion of the material. Such reviews are referred to as partial reviews.

The following applies to partial reviews.

- They may be used if the material to be reviewed has previously been subject to a complete review

- They shall not be used if the material has undergone major changes

- The sections to be reviewed must be stated in the review objectives

- The review meeting must agree that a partial review is appropriate and that the changes do not affect the remainder of the document.

13

B.2. Management review process

B.2.1. Purpose

A management review is the formal evaluation of a project by a designated review team to determine whether the project is proceeding according to the project and quality plans.

During the review meeting the whole review team examines plans, or the progress against applicable plans, standards and/or guidelines. Each problem area is recorded. When critical data and information cannot be provided, an additional meeting shall be scheduled to complete the management review process.

The results of the review are published in a report.

While management reviews can be done at any time, they are normally done at the end of each phase of the project.

The objectives of the management review process are as follows.

- Ensure activities move forward according to plan, based on an evaluation of the product development status.

- Change project direction or to identify the need for alternative planning.

- Maintain overall control of the project through the adequate allocation of resources.

B.2.2. Initiation

B.2.2.1. Initiating event

Management reviews are scheduled activities; their timing is determined by the project and quality plans. Unscheduled management reviews may be called by the project manager, project steering committee or quality manager.

B.2.2.2. Staffing

The review leader for scheduled management reviews is nominated by the project manager, unless specified otherwise in the project or quality plan. The review leader for unscheduled management reviews is determined by the party initiating the review in consultation with the project manager.

A management review shall consist of at least three people. The roles and responsibilities of these people are given below.

Role	Responsibility
Review leader	• Issue the review notification. • Produce and distribute the review material in advance. • Chair the review meeting to ensure its objectives are met. • Issue the review report. • Prepare for the review by studying the distributed material. • Participate in the review to ensure its objectives are met.

Recorder	• Record the action points and decisions during the review. • Reflect any changes to the review objectives and/or agenda in the review report. • Prepare for the review by studying the distributed material. • Participate in the review to ensure its objectives are met.
Reviewer(s)	• Prepare for the review by studying the distributed material. • Participate in the review to ensure its objectives are met

Table 1. Roles and duties of Review group

B.2.2.3. Notification

Participants in a management review shall be given at least five working days notice of the review. The review notification shall include the following information.

- Project name and number.

- Date, time and venue of the review meeting.

- Type of review - (i.e. management, technical, inspection or walkthrough).

- List if input material.

- Statement of review objectives.

- List of issues ensuring objectives cover all issues to be discussed.

- Review team - who will participate and in what role (i.e. moderator/leader, author, recorder etc.).

- Agenda.

- Date, time, venue and expected duration of the preliminary briefing (if required).

Where a management review determines that subsequent meetings are required to complete the review date & time decided by the meeting (notification is not required).

B.2.2.4. Input material

The input material shall be made available to the participants by the review leader at least two working days prior to the review.

The minimum input to the management review is as follows.

- Statement of objectives (in review notification).

- Current project plan, schedule and cost data.

- Reports (i.e. managerial review reports, technical review reports, audit reports) from other reviews/audits which have already been done.

- Reports of resources assigned to the project.

- Data on the documents and software modules completed, or in progress.

- Additional information as required by the review leader.

B.2.2.5. Objectives

The objectives of a management review may vary considerably. An example set of objectives that may be used by a review leader follows.

- Check the project status for compliance with the expected status as outlined in the project plan (i.e. milestones reached on schedule, and that the required documents and software modules exist and have been reviewed and approved).

- Identify any departures from the project plan, making a note of any risks that might apply.

- Determine whether the project is being adversely affected by internal and/or external factors not identified in the project plan.

- Compile a list of issues and recommendations to be addressed by the project steering committee.

- Compile a list of issues and recommendations which will be addressed by other people involved in the project.

- Identify a continuing course of action - expressed as an action point resolving to make certain recommendations.

B.2.3. Conduct

B.2.3.1. Preliminary briefing (optional)

A suitably well-informed project person presents an overview to the review team. This can happen either as part of the examination session or at another time.

B.2.3.2. Agenda

Management reviews shall be conducted using the agenda outlined below. Item 3 shall be expanded to form the detailed agenda.

1. Review the objectives (to allow refinement by the review meeting).

2. Review the agenda (to allow refinement by the review meeting).

3. Review the material relating to the stated objectives and the list of issues, point by point. During the review, the recorder shall accurately note the following.

   - Issues identified
   - Action points
   - Decision list

4. Ensure all objectives have been addressed.

5. Review the decision list.

6. Review the action point list.

7. Review the issues and recommendations list.

8. Make the exit decision

9. Set date for continuation of review (if applicable).

10. Set date of subsequent review (if applicable).

11. Complete the review summary.

Note: Project plan amendments or any product rework resulting from the review is not considered to be part of the management review, except where it is needed as additional input to complete the examination.

B.2.3.3. Exit decision

The exit decision shall be one of the following.

- **Pass** - material accepted as presented.

- **Provisional Pass** - material accepted conditional on validation of rework.

- **Fail** - material to be subject to another review after rework

If the review must be continued a further review meeting shall be scheduled. The review is not complete until one of the above exit decisions is made.

B.2.4. Follow-up actions

The review details and outcomes shall be documented in the review report.

The management review report and the issues & actions/decisions lists shall be forwarded to the steering committee and/or originating party.

B.3. Technical review process

B.3.1. Purpose

A technical review is a formal team evaluation of documents and software modules that can be performed at any time during project development.

The objectives of the technical review process are to evaluate documents, and specific software module(s) and ensure the following.

- The content of the items is correct.

- There is conformance with the specification and standards.

- Identification and categorisation of issues and defects.

- Provision of decisions based on the careful consideration of alternatives.

- The development or maintenance of the software module(s) is being done according to the plans, standards and guidelines applicable to the project.

- Changes to the software module(s) are properly implemented and affect only those system areas identified by the change specification.

Technical review shall be done before approval of project documents and the results of the review shall be published in a report.

Technical reviews apply to both development and maintenance activities.

Distinction between technical review and walkthrough - several important distinctions shall be made between these two otherwise similar activities.

Technical Review	Walkthrough
Mandatory - routine project activity.	Optional project activity.
Focus on finding defects and checking conformance with specification.	Focus on finding more effective ways of meeting the specification..

Table 3. Distinctions between Technical Reviews and Walkthroughs

B.3.2. Initiation

B.3.2.1. Initiating event

Technical reviews are scheduled activities; their timing is determined by the project and quality plans. They are triggered by the completion of specific documents, software modules or project phases. Unscheduled technical reviews may be called by the project manager, project steering committee or quality manager.

B.3.2.2. Staffing

The review leader for technical reviews is nominated by the project manager.

A technical review shall consist of at least three people. The roles and responsibilities of these people are given below.

Role	Responsibility
Review leader	• Issue the review notification. • Chair the review meeting to ensure its objectives are met. • Issue the review report. • Prepare for the review by studying the distributed material. • Participate in the review to ensure its objectives are met.
Recorder	• Record the action points and decisions arising from the meeting. • Reflect any changes to the review objectives and/or agenda in the review report. • Prepare for the review by studying the distributed material. • Participate in the review to ensure its objectives are met.

Reviewer(s)	• Prepare for the review by studying the distributed material. • Participate in the review to ensure its objectives are met
Author	• Participate in the review to ensure its objectives are met. • Produce and distribute the review material in advance. • Can act as recorder or review leader.

Table 4. Roles and responsibilities of the Technical Review

B.3.2.3. Notification

Participants in a technical review shall be given at least two working days notice of the review. The review notification shall include the following information.

- Project name and number.

- Date, time and venue of the review meeting.

- Type of review - (i.e. management, technical, inspection or walkthrough).

- Input material - the document or software module to be reviewed (technical reviews normally focus on a single document, or closely related documents). Materials to be supplied at least one working day before the meeting.

- Statement of review objectives

- Review team - who will participate and in what role (i.e. moderator/leader, author, recorder etc.).

- Agenda.

- Date, time, venue and expected duration of the preliminary briefing (if required).

Where a technical review determines that subsequent meetings are required to complete the review it is not necessary to give two working days notice.

B.3.2.4. Objectives

The objectives of a technical review are as follows. The review leader may include additional objectives as required.

- Check the document, software module or item against the relevant standards, specifications and/or guidelines so as to identify defects.

- Check the technical content of the reviewed material.

- Compile a list of technical issues as well as who will address the respective issues.

- Identify any other relevant issues.

- If proposed as a partial review, it becomes a review objective (i.e. 'It is established that a partial review of (nominated sections) is sufficient').

B.3.3. Conduct

B.3.3.1. Preliminary briefing (optional)

A suitably well-informed project person presents an overview to the review team. This can happen either as part of the examination session or at another time.

B.3.3.2. Agenda

Technical reviews shall be conducted using the broad agenda outlined below. Item 3 shall be expanded to form the detailed agenda.

1. Review the objectives (to allow refinement by the review meeting).

2. Review the agenda (to allows refinement by the review meeting).

3. Review the material relating to the stated objectives and the list of issues, point by point. During the review, the recorder shall accurately note the following.

   - Defects found
   - Action points.
   - Decision list

4. Ensure all objectives have been addressed.

5. Review the decision list.

6. Review the action point list.

7. Review the defects list (optional).

8. Make the exit decision

9. Set date for continuation of review (if applicable).

10. Set date of subsequent review (if applicable).

11. Complete the review summary.

B.3.3.3. Exit decision

The exit decision shall be one of the following.

- **Pass** - material accepted as presented.

- **Provisional Pass** - material accepted conditional on validation of rework.

- **Fail** - material to be subject to another technical review after rework

If the review must be continued a further review meeting shall be scheduled. The review is not complete until one of the above exit decisions is made.

B.3.4. Follow-up actions

The review details and outcomes shall be documented in the technical review report.

The technical review report and the issues/actions and decision lists are forwarded to the project manager.

B.4. Walkthrough process

B.4.1. Purpose

During the walkthrough meeting, the author makes an overview presentation of the software modules and/or documents under consideration. This is followed by a general discussion by the participants after which the presenter 'walks through' the software module and/or document in detail.

As the walkthrough progresses, suggested changes and improvements are recorded as well as areas that have the potential to derail the project.

When the walkthrough is finished, the working notes are consolidated into one report.

The objective of a walkthrough is to evaluate a software module to find ways of improving the software module(s) and to consider more efficient ways of meeting the specification. Although walkthroughs are traditionally associated with code examinations, this process is also applicable to other software modules and documents (i.e. architectural design, detailed design, test plans/procedures and change control procedures).

Walkthroughs help to highlight areas whose efficiency or readability could be improved and also to show the participants other ways of doing things.

Walkthroughs are held at any time in the project life cycle and apply to both development and maintenance activities.

Distinction between walkthrough and technical review - several important distinctions can be made between these two activities.

B.4.2. Initiation

B.4.2.1. Initiating event

Walkthroughs are an optional review which can be scheduled in the project and quality plans but is more often called at short notice as required.

A walkthrough is convened when the following conditions are met.

- The project manager authorises it.

- The author of the software module(s) considers that it would assist in the development of higher quality deliverables.

- The leader is appointed and has concluded that the walkthrough is warranted and sufficient information is available to enable it to occur.

B.4.2.2. Staffing

The walkthrough leader is nominated by the project manager.

A walkthrough shall consist of at least three people. The roles and responsibilities of these people are given below.

Role	Responsibility
Walkthrough leader	• Issue the walkthrough notification. • Chair the walkthrough meeting to ensure its objectives are met. • Prepare for the walkthrough by studying the distributed material. • Participate in the walkthrough to ensure its objectives are met.
Recorder	• Record the action points, issues and decisions arising from the meeting. • Issue the walkthrough report. • Reflect any changes to the walkthrough objectives and/or agenda in the walkthrough report. • Prepare for the walkthrough by studying the distributed material. • Participate in the walkthrough to ensure its objectives are met.
Reviewer(s)	• Prepare for the walkthrough by studying the distributed material. • Participate in the walkthrough to ensure its objectives are met
Author	• Produce and distribute the walkthrough material in advance. • Contribute to the walkthrough through their special understanding of the software modules. • Can act as recorder or walkthrough leader.

Table 5. Roles and responsibilities in Walkthroughs

B.4.2.3. Notification

Participants in a walkthrough shall be given at least one working days notice of the meeting. The notification shall include the following information.

- Project name and number.

- Date, time and venue of the review meeting.

- Type of review - (i.e. management, technical, inspection or walkthrough).

- Input material - the document, software module or other item to be reviewed. Walkthroughs normally focus on a single document, or closely related documents. Materials should be supplied at least one working day before the meeting.

- Statement of review objectives.

- Review team - who will participate and in what role (i.e. moderator/leader, author, recorder etc.).

- Agenda.

- Date, time, venue and expected duration of the preliminary briefing (if required).

Where a walkthrough determines that subsequent meetings are required to complete the walkthrough it is not necessary to give five working days notice.

B.4.2.4. Objectives

The objectives of a walkthrough are as follows. The walkthrough leader may include additional objectives as required.

- Walk through the document, review material or software module(s) and raise issues of concern or identify areas for improvement.

- Identify areas that have the potential to derail the project.

- Identify alternative approaches if appropriate.

- Record all comments and decisions for inclusion in the walkthrough report.

B.4.3. Conduct

B.4.3.1. Preliminary briefing (optional)

The author presents an overview of the software module(s).

B.4.3.2. Agenda

Walkthroughs shall be conducted using the broad agenda outlined below. Item 3 shall be expanded to form the detailed agenda.

1. Review the objectives (to allow refinement by the walkthrough meeting).

2. Review the agenda (to allows refinement by the walkthrough meeting).

3. Review the material relating to the stated objectives and the list of issues. During the walkthrough, the recorder shall accurately note the following.

    -   Review issues
    -   Action points
    -   Decisions

4. Ensure all objectives have been addressed.

5. Review the decision list.

6. Review the action point list.

7. Review the issues list (optional).

8. Make the exit decision.

9. Set date for continuation of walkthrough (if applicable).

10. Set date of subsequent walkthrough (if applicable).

11. Complete the walkthrough summary.

B.4.3.3. Exit decision

The exit decision shall be one of the following.

-   Walkthrough complete.

-   Subsequent walkthrough recommended.

B.4.4. Follow-up actions

The walkthrough details and outcomes shall be documented in the walkthrough report.

The walkthrough report and the issues & actions/decisions lists are forwarded to the project manager.

B.5. Audit process

B.5.1. Purpose

Audits are done according to a prepared audit plan. The audit plan establishes a procedure to conduct the audit and for follow-up action on the audit findings. The audit plan is prepared by the audit leader or auditing organisation.

Audits apply to the following.

- Quality management system modules.

- Development/maintenance projects.

- Small projects.

In performing the audit, audit personnel evaluate the input material and processes identified in the audit plan for compliance with the audit criteria.

The results of the audit are documented and are submitted to management. The report includes a list of items in non-compliance or other issues for the subsequent review and action. Where appropriate, recommendations are appended to the audit results. The report is ultimately placed in the project file for auditability.

The objective of a software quality audit is to confirm that products and processes are being developed in keeping with the prescribed standards, guidelines, specifications and procedures.

B.5.2. Initiation

B.5.2.1. Initiating event

An audit is convened when at least one of the following events occur.

- A special project milestone is reached and the audit is initiated according to the project plan.

- An external party (i.e. the customer) requires an audit at a particular project milestone or date. This requirement shall already be in the project agreement.

- The project manager, quality manager, or steering committee requests the audit after establishing a clear need for it.

B.5.2.2. Staffing

The audit leader for scheduled audits is nominated by the quality manager, unless specified otherwise in the project plan. The leader for unscheduled audits is determined by the party initiating the audit in consultation with the quality manager.

An audit shall be conducted by an audit team, lead by the Audit Leader, on behalf of the Quality Manager. The roles and responsibilities are given below.

Role	Responsibility
Quality manager	• Convenes the audit according to a prepared schedule encompassing all internal and external projects. • Is not necessarily involved in audit.
Audit leader	• Issue the audit notification. • Produce and distribute the audit plan in advance. • Manage the audit to ensure its objectives are met. • Issue the audit report. • Prepare and issue the audit plan.

Table 6. Roles and responsibilities in Audits

B.5.2.3. Preparation

The audit team prepares for the audit by doing the following.

- Understand the organisation - it is essential to identify functions and activities performed by the auditee and to identify functional responsibility.

- Understand the products and processes - it is a prerequisite for the team to learn about the products and processes being audited through readings and briefings.

- Develop an approved audit plan.

- Prepare for the audit report - choose the administrative reporting mechanism to be used throughout the audit process.

In addition, the audit team leader makes the necessary arrangements for the following.

- Audit team orientation and training, as required.

- Facilities for audit interviews.

- Materials, documents and tools needed by the audit process.

- The software module(s) to be audited (i.e. documents, computer files, personnel to be interviewed.

- The preliminary audit briefing (if required).

B.5.2.4. Notification

The project manager of the project being audited shall be given at least fifteen working days notice of the commencement of the audit. If the audit is specified in the project agreement the audit plan shall be prepared and approved before the signing of the project agreement which shall contain specific reference to it. The notification shall include the following information.

- Project name and number.

- Date, time, venue and expected duration of the audit.

- Type of review - (i.e. audit).

- Date, time, venue and expected duration of the preliminary briefing (if required).

- A copy of the approved audit plan

B.5.3. Conduct

B.5.3.1. Preliminary briefing (optional)

A preliminary briefing of the auditee is optional but highly recommended to begin the examination phase of the audit. The overview session, led by the audit leader, provides the following.

- Overview of existing agreements.

- Overview of product and processes being audited.

- Overview of the audit process, its objectives and follow-up actions.

- Expected contribution to the audit process by the auditee (i.e. the number of people to be interviewed, meeting facilities etc.)

- Specific audit schedule.

B.5.3.2. Audit

The Audit shall be conducted in accordance with the approved audit plan.

B.5.3.3. Exit criteria

The audit is complete when the following steps have been completed.

1. Each module within the scope of the audit has been examined.

2. Findings have been presented to the auditee.

3. The auditees response to the draft findings have been received and evaluated.

4. Final findings have been formally presented to the auditee and initiating entity.

5. Audit report has been prepared and submitted to recipients designated in the audit plan.

6. The recommendations report (if applicable) has been prepared and submitted to recipients designated in the audit plan.

7. All of the auditors follow-up actions included in the scope (or customer agreement) of the audit has been performed.

B.5.4. Follow-up actions

B.5.4.1. Overview

The findings of the audit review are documented in the audit report. A draft copy of the report is sent to the auditee for review/comment within three days of the audit. The auditee returns the draft report to the auditor within three days of receipt.

The audit team conducts a post-audit conference with the auditee as a necessary part of the process of producing the final audit report. The final report is then issued by the audit leader to the parties nominated in the project and quality plans, or the initiation of the audit if applicable.

B.5.4.2. Audit report

The audit details and outcomes shall be documented in the audit report.

The audit report shall include the following information.

- Audit Identification
- Report title.
- Auditing organisation.
- Audited organisation.
- Date of audit.
- Scope - make reference to the audit plan which details the scope and the objectives of the audit. These should

not be restated in the audit report unless appreciable changes have been made to either the scope of the audits.

▪ Conclusions - a summary and interpretation of the audit findings, including the key items of non-conformance.

▪ Synopsis - a list of all the audited software elements, processes and associated findings.

▪ Follow-up - the type and timing of audit follow-up activities.

▪ Cost information - indicate the actual cost of the audit, breaking it down into auditor and auditee categories.
  – Audit preparation.
  – Conduct of audit.
  – Preparation of audit report.

▪ Recommendations - when stipulated by the audit plan, recommendations shall be provided to the audited organisation or the entity that initiated the audit. Recommendations shall be reported separately from results.

▪ Authorisation - signature, title, division, branch.

B.6. Inspection process

B.6.1. Purpose

Inspections are rigorous formal evaluations conducted by peers and designed to identify defects in the inspected material. An inspection team is normally comprised of three to six participants. The team leader is an impartial person independent of the development project.

The resolution of defects is mandatory and any rework performed is formally verified. The defect data is systematically collected and recorded. When the inspection data is analysed, steps can be identified which help to improve both the product and the development process. The results of the inspection are published in a report which is later placed in the project file for auditability.

An inspection is a formal peer examination that does the following.

- Verifies that the inspected material satisfies its specifications.

- Verifies that the inspected material conforms to the applicable standards.

- Identifies deviation from standards and specifications.

- Collects software engineering data.

- Does not examine alternatives or stylistic issues.

42

B.6.2. Initiation

B.6.2.1. Initiating event

Inspections are an optional activity which is seldom performed. The timing of planned inspections is determined by the project and quality plans if specified. Unplanned inspections may be called by external entities like the quality manager or customer.

Prior to conducting an inspection it shall be determined that the software modules and documents and other documentation are complete enough to be able to support the inspection objectives.

B.6.2.2. Staffing

The review leader for scheduled inspections is nominated by the quality manager. The review leader for unscheduled inspections is determined by the party initiating the inspection in consultation with the quality manager.

An inspection shall consist of at least four people. The roles and responsibilities of these people are given below.

Role	Responsibility
Inspection leader (external person)	• Issue the inspection notification. • Distribute the inspection material in advance. • Chair the inspection meeting to ensure its objectives are met. • Issue the inspection report. • Prepare for the inspection by studying the distributed material. • Participate in the inspection to ensure its objectives are met. • Shall be an external person (i.e. quality manager). • May act as recorder.
Reader	• Lead the inspection team through the software modules in a thorough and logical manner. • Prepare for the inspection by studying the distributed material. • Participate in the inspection to ensure its objectives are met.
Recorder	• Record the defects, inconsistencies, omissions and ambiguities revealed by the inspection. • Record the action points and decisions arising from the meeting. • Reflect any changes to the inspection objectives and/or agenda in the inspection report. • Prepare for the inspection by studying the distributed material. • Participate in the inspection to ensure its objectives are met.

Inspector(s) (shall not be project team member)	• Prepare for the inspection by studying the distributed material. • Identify and describe defects, inconsistencies, omissions and ambiguities in the software modules. • Represent pertinent viewpoints (i.e. requirements, design, code, test, project management, quality management) as required by the inspection.
Author	• Contribute to the inspection through their special understanding of the software modules. • Shall not perform any other role in the inspection process.

Table 6. Roles and responsibilities in Inspections

B.6.2.3. Notification

Participants in an inspection shall be given at least five working days notice of the inspection. The inspection notification shall include the following information.

- Project name and number.

- Date, time and venue of the review meeting.

- Type of review - (i.e. inspection).

- Input material (i.e. documents and/or software modules).

  - Inspections are done on single documents, software modules or other items, or on a set of closely related groups thereof.
  - Inspections can only be done on 'approved' documents, software modules or other items.

- Materials should be supplied at least 2 working days before the meeting.
- Statement of review objectives.

- Inspection team - who will participate and in what role (i.e. moderator/leader, author, recorder etc.).

- Agenda.

- Date, time, venue and expected duration of the preliminary briefing (if required).

Where an inspection determines that subsequent meetings are required to complete the inspection it is not necessary to give five working days notice.

B.6.2.4. Objectives

The objectives of an inspection are as follows. The review leader may include additional objectives as required.

- To assess the input material for compliance against the applicable standards.

- To assess the input material for compliance against the specification (if applicable).

- To focus on fault-detection rather than solution-hunting.

- To consider the material objectively, systematically developing the defect list.

- Review the defect list to confirm that it is accurate and complete.

- To reach an exit decision.

B.6.3. Conduct

B.6.3.1. Preliminary briefing (optional)

If the inspection leader considers it necessary, the author presents an overview of the material being inspected.

B.6.3.2. Agenda

Inspections shall be conducted using the broad agenda outlined below. Item 3 shall be expanded to form the detailed agenda.

1. Review the objectives (to allow refinement by the inspection meeting).
2. Review the agenda (to allows refinement by the inspection meeting).
3. Review the material relating to the stated objectives and the list of issues. During the inspection, the recorder shall accurately note the following.

    – Defects.

    – Action points.

    – Decisions made and their reasons.
1. Ensure all objectives have been addressed.
2. Review the decision list.
3. Review the action point list.
4. Review the defect list.

5. Make the exit decision.

6. Set date for continuation of inspection (if applicable).

7. Set date of subsequent inspection (if applicable).

8. Complete the inspection summary.

B.6.3.3. Exit decision

The exit decision shall be one of the following.

- **Pass** - material accepted as presented.

- **Provisional Pass** - material accepted conditional on validation of rework.

- **Fail** - material to be subject to another inspection after rework

If the inspection must be continued a further meeting shall be scheduled. The inspection is not complete until one of the above exit decisions is made.

B.6.4. Follow-up actions

The inspection details and outcomes shall be documented in the inspection report.

The inspection report and the issues & actions/decisions lists are forwarded to the initiator of the inspection and the project manager.

B.7. Review follow-up actions

Note: Review follow-up actions identify the required output from the various processes that *explicitly make reference to this section.*

The output from a review comprise the following material. Together they are treated as the 'review report'.

- Review Notification / Agenda.

- Review Summary Report.

- Review Issue.

- Defects List.

- Review Action List.

- Review Decision List.

- Other information as required.

B.7.1. Review notification

The review notification shall include the following information.

- Project name and number.

- Date, time and venue of the review meeting.

- Type of review - (i.e. management, technical, inspection or walkthrough).

- Input material (i.e. ID and version of documents and/or software modules).

- Statement of review objectives

- Review team - who will participate and in what role (i.e. moderator/leader, author, recorder etc.).

- Agenda.

- Date, time, venue and expected duration of the preliminary briefing (if required).

B.7.2. Review summary

The review summary report shall include the following information.

- Project name and number.
- Review date, time and venue.
  - Meeting commencement.
  - Meeting completion.
  - If multiple sessions, give start and finish times and venue for each session.
- Date and time of review meeting completion.
- Type of review - (i.e. management, technical, inspection or walkthrough).
- Review input list

- Any modifications to the review objectives from those stated in the review notification.

- Any modifications to the review agenda from those stated in the review notification.

- Review team - those actually present for the review and their role.

- Metrics - indicating the following detail.

  - Preparation Effort
    The total time spent by all participants in preparation for the review.

  - Meeting Effort
    Total time spent in the actual review sessions (number of people by total duration of the meeting sessions).

  - Total review effort (preparation + meetings).

  - Number of major issues/defects.

  - Number of minor issues/defects (optional).

  - Number of action points.

  - Number of decisions.

  - Size of material reviewed - for documents, indicate the number of pages; for software source, the number of lines, including comments (not applicable to management reviews and audits).

- Exit decision - the exit decision shall be one of the following.

- **Pass** - material accepted as presented.

- **Provisional Pass** - material accepted conditional on validation of rework.

- **Fail** - material to be subject to another inspection after rework

If the review must be continued a further review meeting shall be scheduled. The review is not complete until one of the above exit decisions is made.

B.7.3. Review issues

Issues arise only from management reviews and walkthroughs (defects arise from technical reviews and inspections).

The review issues list shall include the following information for each issue identified.

- Issue/defect number.

- Reference to the document or item in which issue/defect was found.

- Issue/defect description.

- Issue/defect type - optional (refer appendix I)

- Issue/defect class - optional but recommended (refer appendix I)

- Issue/defect severity - mandatory (refer appendix I)

- Person responsible for addressing the issue/defect?

- Name or signature of the person addressing the issue.

Note: When reviewing a document or software module it is common to note the issues or defects and corrections on a printed copy of the document. The marked-up copy shall be prominently labelled 'Review Master' and is included in the review report. Where the review master is included in the review report, all major (severity) issues must be recorded in the review issues list.

B.7.4. Defects list

The defects list uses the same form as the issues list. Defects only arise from technical reviews and inspections, (while issues come from management reviews and walkthroughs).

B.7.5. Review action point list

Action points are items arising from the meeting which someone is nominated to perform. Action points are generally not related to the correction of issues or defects. It is implicitly understood that action points arising from document reviews shall be performed by the author; they are not explicitly stated as an action point.

The review action point list shall include the following information.

- Action point number (sequential number uniquely identifying each point).

- Reference to the document or item requiring action.

- Action point description

- Who will implement the Decision (if applicable)

- By what date will the Decision be implemented (if applicable)

- Name or signature of the person effecting the decision.

- Date completed (Date of signature).

B.7.6. Review decision list

Decisions are a clear choice made by the meeting between two or more alternatives and are not actions in themselves. If a decision requires specific actions to be performed, the action points must be listed in the action list.

The review decision list shall include the following information.

- Decision number (sequential number uniquely identifying each point).

- Reference to the document or item in requiring a decision.

- Decision description

▪ Reason - specify the significant factor determining the decision (for later reference).

B.8. Audit plan

An audit plan is developed for each audit. The following diagram specifies the minimum contents of the plan.

The audit plan shall be approved by the quality manager before being given to the project manager as part of the audit notification. Discussions between the audit leader and the project manager can be held before notification is sent.

1. Introduction
    1.1 Background
    1.2 Objectives
    1.3 Scope
    1.4 Definitions, acronyms and abbreviations
    1.5 References
2. Project processes
3. Software module(s)
4. Reports
5. Report distribution
6. Follow-up activities
7. Requirements
8. Objective audit criteria
9. Audit procedures and checklists
10. Audit personnel
11. Organisations involved
12. Overview session

13. Audit Budget

Table 7: Audit plan table of contents

B.8.1. Introduction (section 1)

The purpose of this section of the audit plan is to provide an overview of the purpose and scope of the proposed audit.

B.8.1.1. Background (section 1.1)

Include a brief description of the circumstances surrounding the audit.

B.8.1.2.   Objectives (section 1.2)

Specify the audit objectives.

B.8.1.3. Scope (section 1.3)

The scope identifies the following.

- Specific audit concerns.

- What the audit will and will not address.

- The documents and/or software modules being audited.

B.8.1.4. Definitions, acronyms and abbreviations (section 1.4)

Include all definitions etc. needed to understand the audit plan or which would be helpful in understanding the plan. This information can be provided by referring to appendix(es) or to other documents.

B.8.1.5. References (section 1.5)

List the documents referenced in some way by the plan. The documents shall already exist - that is, reference shouldn't be made to documents planned to be written in the future.

This information can be provided by referring to appendix(es) or to other documents.

B.8.2. Project processes (section 2)

Specify the project processes to be examined (input) and the time frame for the audit team review.

B.8.3. Software module(s) (section 3)

Specify the software module(s) to be reviewed (input) and their availability. Where sampling is used, a statistically valid sampling method should be used to establish selection criteria and sample size.

B.8.4. Reports (section 4)

Identify the results report and an optional recommendations report and their format. If recommendations are required, the audit plan should specifically say so.

B.8.5. Report distribution (section 5)

Specify the required report distribution.

B.8.6. Follow-up activities (section 6)

Specify the required follow-up activities.

B.8.7. Requirements (section 7)

Specify the requirements, necessary activities, modules and procedures to meet the scope of the audit.

B.8.8. Objective audit criteria (section 8)

Specify the objective audit criteria to provide the basis for determining compliance (input). Audit criteria shall be given for all processes and software elements identified in sections 2 and 3.

B.8.9. Audit procedures and checklists (section 9)

Specify the audit procedures and checklists.

B.8.10. Audit personnel (section 10)

Specify the audit personnel involved, including how many, what skills they have, a description of their background and responsibilities.

B.8.11. Organisations involved (section 11)

Specify the auditor and the auditee.

B.8.12. Overview session (section 12)

Specify the date, time, place, agenda and intended audience of the preliminary audit briefing session.

B.8.13. Audit Budget (section 13)

This section specified the budget for the audit and identifies the internal amount to which the audit expenses and labour cost for both audit team and project staff shall be charged.

C. Appendix I Issue type/ class/ severity

C.1. Item types (optional)

Code	Type Name	Description	Usage
M	Maintainability	Used for issues/defects which refer to things which make the material difficult to maintain	Could be used for any type of review. Would generally be used with classifications of Inconsistent, or Wrong
SCH	Schedule Compliance	Used for issues/defects which detail non-compliance to schedule for the task.	Only for management reviews
SC	Scope Compliance	Used for issues/defects which detail non-compliance to the scope of the task.	Only for management reviews
SP	Specification Compliance	Used for issues/defects which detail non-compliance to specifications. The specification may be written, verbal, or an objective the material is to fulfil.	Could be used for any type of review. Generally used with classifications of Incomplete, Missing, or Wrong
SG	Spelling and Grammar	Used in relation to documents where the issues/defect refers to spelling and grammatical errors.	Could be used for any type of review. Would generally be used with the classification of Wrong.
ST	Standards Compliance	Used for issues/defects which detail non-compliance to relevant standards	Could be used for any type of review. Generally used with classifications of Incomplete, Missing or Wrong

SS	Style and Structure	Used for issues/defects which require changes to the style or structure of the material	Could be used for any type of review. Would generally be used with classifications of Inconsistent, or Wrong
UI	Unresolved Issue	Used for issues/defects which are issues to be resolved outside the walk-through, or are to be left unresolved and associated risk accepted.	Would generally be used for formal walk-throughs to describe issues which are to be accepted as risk. In this case it would be used with the Risk classification.
WO	Wording	Used in relation to documents where the issues/defect refers to changes to wording only.	This type is not to be used for issues/defects which require wording changes to be made in order to satisfy specifications or standards. In these cases SP or ST are to be used. Can be used with all except Risk.

Table 8. Item/class (optional)

C.2. Item classifications (optional but recommended)

Code	Classification	Description	Usage
AM	Ambiguous	Used for issues/defects which indicate that areas of the material are unclear	Generally used with issues/defects of type WO.
IN	Incomplete	Used for issues/defects which indicate that areas of the material are not complete	Generally used with issues/defects of type SP, ST, or WO
IC	Inconsistent	Used for issues/defects which indicate inconsistencies within the material	Generally used with issues/defects of type SS or WO.

62

MI	Missing	Used for issues/defects which indicate that areas of the material are missing	Generally used with issues/defects of type SP, ST, or WO
RI	Risk	Used for issues/defects which will cause risk	Generally used with issues/defects of type UI
WR	Wrong	Used for issues/defects which indicate that areas of the material are not correct	Generally used with issues/defects of type M, SP, SG, ST, SS, or WO

Table 9. Item classifications (optional but recommended)

C.3. Item severities (mandatory)

Code	Severity	Description	Usage
Minor	Minor	Defects that would affect only the non functional aspects of the material. (This definition is from IEEE 1028-1988 section 6.10.4.)	Can be used with issues/defects of any type and classification.
Major	Major	Defects that would result in failure of the material or an observable departure from specification. (This definition is from IEEE 1028-1988 section 6.10.4.)	Can be used with issues/defects of any type and classification.

Table 10. Item severities (mandatory)

www.ingramcontent.com/pod-product-compliance
Lightning Source LLC
Chambersburg PA
CBHW071031050326
40689CB00014B/3610